Macmillan/McGraw-Hill Science

Sound
AND Light

AUTHORS

Mary Atwater
The University of Georgia
Prentice Baptiste
University of Houston
Lucy Daniel
Rutherford County Schools
Jay Hackett
University of Northern Colorado
Richard Moyer
University of Michigan, Dearborn
Carol Takemoto
Los Angeles Unified School District
Nancy Wilson
Sacramento Unified School District

See the flash, hear the boom

Macmillan/McGraw-Hill School Publishing Company
New York   Chicago   Columbus

MACMILLAN / McGRAW-HILL

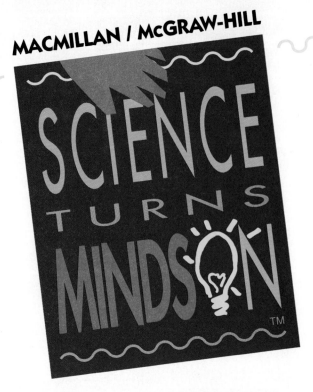

SCIENCE TURNS MINDS ON™

CONSULTANTS

Assessment:

Janice M. Camplin
Curriculum Coordinator, Elementary Science
Mentor, Western New York
Lake Shore Central Schools
Angola, NY

Mary Hamm
Associate Professor
Department of Elementary Education
San Francisco State University
San Francisco, CA

Cognitive Development:

Dr. Elisabeth Charron
Assistant Professor of Science Education
Montana State University
Bozeman, MT

Sue Teele
Director of Education Extension
University of California, Riverside
Riverside, CA

Cooperative Learning:

Harold Pratt
Executive Director of Curriculum
Jefferson County Public Schools
Golden, CO

Earth Science:

Thomas A. Davies
Research Scientist
The University of Texas
Austin, TX

David G. Futch
Associate Professor of Biology
San Diego State University
San Diego, CA

Dr. Shadia Rifai Habbal
Harvard-Smithsonian Center for Astrophysics
Cambridge, MA

Tom Murphree, Ph.D.
Global Systems Studies
Monterey, CA

Suzanne O'Connell
Assistant Professor
Wesleyan University
Middletown, CT

Environmental Education:

Cheryl Charles, Ph.D.
Executive Director
Project Wild
Boulder, CO

Gifted:

Sandra N. Kaplan
Associate Director, National/State Leadership
Training Institute on the Gifted/Talented
Ventura County Superintendent of Schools Office
Northridge, CA

Global Education:

M. Eugene Gilliom
Professor of Social Studies and Global Education
The Ohio State University
Columbus, OH

Merry M. Merryfield
Assistant Professor of Social Studies and Global Education
The Ohio State University
Columbus, OH

Intermediate Specialist

Sharon L. Strating
Missouri State Teacher of the Year
Northwest Missouri State University
Marysville, MO

Life Science:

Carl D. Barrentine
Associate Professor of Biology
California State University
Bakersfield, CA

V.L. Holland
Professor and Chair, Biological Sciences Department
California Polytechnic State University
San Luis Obispo, CA

Donald C. Lisowy
Education Specialist
New York, NY

Dan B. Walker
Associate Dean for Science Education and Professor of Biology
San Jose State University
San Jose, CA

Literature:

Dr. Donna E. Norton
Texas A&M University
College Station, TX

Tina Thoburn, Ed.D.
President
Thoburn Educational Enterprises, Inc.
Ligonier, PA

Macmillan/McGraw-Hill School Division
10 Union Square East
New York, New York 10003

Printed in the United States of America

ISBN 0-02-274272-7 / 5

2 3 4 5 6 7 8 9 VHJ 99 98 97 96 95 94 93 92

Laser game

2

Mathematics:

Martin L. Johnson
Professor, Mathematics Education
University of Maryland at College Park
College Park, MD

Physical Science:

Max Diem, Ph.D.
Professor of Chemistry
City University of New York, Hunter College
New York, NY

Gretchen M. Gillis
Geologist
Maxus Exploration Company
Dallas, TX

Wendell H. Potter
Associate Professor of Physics
Department of Physics
University of California, Davis
Davis, CA

Claudia K. Viehland
Educational Consultant, Chemist
Sigma Chemical Company
St. Louis, MO

Reading:

Jean Wallace Gillet
Reading Teacher
Charlottesville Public Schools
Charlottesville, VA

Charles Temple, Ph.D.
Associate Professor of Education
Hobart and William Smith Colleges
Geneva, NY

Safety:

Janice Sutkus
Program Manager: Education
National Safety Council
Chicago, IL

Science Technology and Society (STS):

William C. Kyle, Jr.
Director, School Mathematics and Science Center
Purdue University
West Lafayette, IN

Social Studies:

Mary A. McFarland
Instructional Coordinator of
Social Studies, K-12, and
Director of Staff Development
Parkway School District
St. Louis, MO

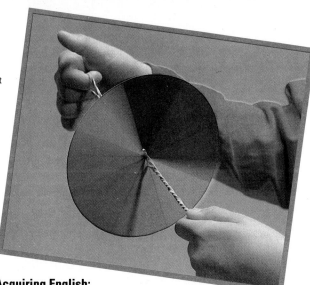

Students Acquiring English:

Mrs. Bronwyn G. Frederick, M.A.
Bilingual Teacher
Pomona Unified School District
Pomona, CA

Misconceptions:

Dr. Charles W. Anderson
Michigan State University
East Lansing, MI

Dr. Edward L. Smith
Michigan State University
East Lansing, MI

Multicultural:

Bernard L. Charles
Senior Vice President
Quality Education for Minorities Network
Washington, DC

Cheryl Willis Hudson
Graphic Designer and Publishing Consultant
Part Owner and Publisher, Just Us Books, Inc.
Orange, NJ

Paul B. Janeczko
Poet
Hebron, MA

James R. Murphy
Math Teacher
La Guardia High School
New York, NY

Ramon L. Santiago
Professor of Education and Director of ESL
Lehman College, City University of New York
Bronx, NY

Clifford E. Trafzer
Professor and Chair, Ethnic Studies
University of California, Riverside
Riverside, CA

STUDENT ACTIVITY TESTERS

Jennifer Kildow
Brooke Straub
Cassie Zistl
Betsy McKeown
Seth McLaughlin
Max Berry
Wayne Henderson

FIELD TEST TEACHERS

Sharon Ervin
San Pablo Elementary School
Jacksonville, FL

Michelle Gallaway
Indianapolis Public School #44
Indianapolis, IN

Kathryn Gallman
#7 School
Rochester, NY

Karla McBride
#44 School
Rochester, NY

Diane Pease
Leopold Elementary
Madison, WI

Kathy Perez
Martin Luther King Elementary
Jacksonville, FL

Ralph Stamler
Thoreau School
Madison, WI

Joanne Stern
Hilltop Elementary School
Glen Burnie, MD

Janet Young
Indianapolis Public School #90
Indianapolis, IN

CONTRIBUTING WRITER

Don Schaub

SOUND AND LIGHT

Activities!

EXPLORE

TRY THIS

Features

Links

Literature Links

Health Links

Social Studies Link

Math Link

Art Links

CAREERS

SCIENCE TECHNOLOGY and Society

Focus on Environment

Focus on Technology

GLOBAL VIEW

Departments

Off in the distance, you
see a flash of lightning.

Sound & LIGHT

Sound and light—you can't hold them.
They have no substance. They have no
weight. But they're all around you.
What causes them? How do they move
from place to place?

A second later, you hear a loud crash of thunder. The noise is so loud it rattles windows and shakes the house.

Why didn't you hear the thunder when you saw the lightning?

Sound and light are common aspects of the environment. Where would you be without them? Close your eyes and try to imagine what it would be like if light were removed from your life. Put your hands over your ears and you'll have some idea of what it would be like if you could not hear the sounds of everyday life.

Minds On! Think about how much you rely on light and sound each day. Make a list in your *Activity Log* on page 1 of ways that you use sound and light. ●

When we think of sound and light, communication is probably the first thing that comes to mind. But that's only part of the story. Sound and light help us learn, work at jobs, and even protect ourselves from danger. Sound and light provide us with many kinds of entertainment. What kinds of things do you enjoy that rely on sound or light? How do you think sound and light link us with the rest of the world?

 Health Link

Warning Lights

Your ability to hear sound and see light helps protect you from many of the dangers that are part of everyday life.

For example, when an automobile approaches an intersection or a railroad crossing, red warning lights tell the driver to stop. You know not to eat things that are labeled "poison" or to touch things labeled "hot." Being able to see a "Danger" sign keeps you away from areas where you could be injured.

The sound of the smoke detector has awakened many people in time to escape from burning buildings. Tornado sirens and television announcements give us warnings during dangerous weather conditions.

The scream of sirens and the flash of red lights are signals for cars to move out of the way so emergency vehicles can pass.

Minds On!

Because our minds are so busy, we sometimes ignore things going on around us. Many sounds, for example, go unnoticed because our attention is focused on other things. Close your eyes and concentrate on all the sounds you can identify. How many things can you hear that you didn't notice when your eyes were open? ●

Even though sound and light are basic parts of our lives, they may seem mysterious to most of us. You won't learn everything about sound and light in this unit. But some of the mystery will be lifted as you examine just what light and sound are, discover some of the characteristics of each, and explore how we actually hear sound and see light.

What would happen if Garrett Morgan hadn't invented the automatic stop light?

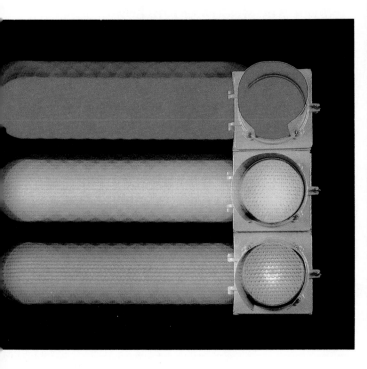

Science in Literature

Life on Earth would be very different without sound and light. Here are lots of books that can tell you more about sound and light.

Rainbows to Lasers by
Kathryn Whyman.
New York: Gloucester Press, 1990.
Light makes the world bright and colorful! It helps you see sizes, shapes, and colors. Your eyes use light to see and you look at things through special instruments, such as microscopes and telescopes.
Rainbows to Lasers will help you find out why shadows form, how rainbows appear in the sky, and how mirrors reflect light.

HANDS · ON · SCIENCE

RAINBOWS TO LASERS

PROJECTS WITH LIGHT

Sound Waves to Music by Neil Ardley. New York: Gloucester Press, 1990.

Sounds are everywhere. Even on a quiet night you will hear them. There are natural sounds made by animals, trees, people, or the wind, and there are artificial sounds made by machines or musical instruments. Read *Sound Waves to Music* to find out about what sound is, how sound moves, and how we hear and use sound.

Other Good Books To Read

Fun With Science—Sound by Terry Cash. New York: Warwick Press, 1987.

This activity-filled book is divided into two sections. In the first section you can discover how sounds are made and how we hear them. In the second section, you can explore musical instruments that produce sounds.

The Seeing Summer by Jeannette Eyerly. New York: J. B. Lippencott, 1981.

In this book a young girl named Carey meets a new neighbor, Jenny, who is blind. When the girls wind up in the hands of kidnappers, they use their resourcefulness to survive a scary adventure.

Fun With Science—Light by Brenda Walpole. New York: Warwick Press, 1987.

This book is filled with experiments you can do to help you answer some questions about light and color.

Follow My Leader by James Garfield. New York: The Viking Press, 1957.

When the firecracker one of his buddies tossed accidently explodes in Jimmy's face, everyone is stunned. Because of this accident, Jimmy will never see again. Read *Follow My Leader* to see how Jimmy learned Braille, had his guide dog trained, and became independent again after the accident.

Sounds are everywhere. Some come from nature and others come from machines. Many are made by people at work or play. We wish some things didn't make sounds. Other things are designed to make sounds. What living and nonliving things make enjoyable sounds? Which ones make unpleasant noises? What causes sound and how does it reach you?

WHAT IS SOUND?

You turn on your radio and adjust the sound level. If you could take your radio into a room without any air in it, you could turn it up full volume and you wouldn't hear a single note! Why is that?

Airplane coming in for a landing

A common source of sounds you hear is people's voices. Have you spoken to anyone today? Has anyone spoken to you? Think about how you use sound to communicate. What kinds of sounds can you produce this very moment?

Minds On! In the introduction to this unit, you made a list in your *Activity Log* on page 1 of how you use sound every day. Go back to the list now and add the possible sources of sounds to it. Think first about what is making the sound. Then try to determine what is really happening to produce that sound. ●

Activity!

How Does Sound Happen?

How many different ways are there to produce sounds? In this activity, you'll investigate what causes sounds. To do this, you'll make a series of different sounds and try to determine how they are all alike and all different.

What You Need

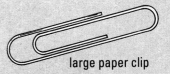

large paper clip

Activity Log pages 2-3

rubber bands of various lengths and thicknesses

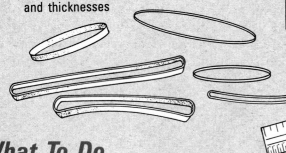

ruler

What To Do

1 Put your goggles on and stretch 1 of the rubber bands and have your partner pluck it. What do you observe? Look carefully! Write your observations in your **Activity Log**.
Safety Tip: Use caution when working with rubber bands.

2 Repeat step 1 using different rubber bands and record your new observations in your **Activity Log**.

Safety!

See the *Safety Tip* in step 1.

3 Bend the paper clip and hold it on the table as shown in the picture.

4 Pluck the upper end of the paper clip and observe what happens. What can you see? Record your observations of the paper clip in your **Activity Log**.

5 Hold the ruler firmly against your desk top with about half of it sticking out over the edge. Pluck the free end of the ruler. Record your observations of the ruler in your **Activity Log**.

What Happened?

1. How would you describe the sounds you made?

2. What other observations did you make about each of the sound-makers?

What Now?

1. What did all the objects have in common while they were producing sounds?

2. What other objects can you think of that produce sound in the same way that the rubber band, paper clip, and ruler did?

3. What did you have to do to make the objects produce sound?

4. How do you think each sound traveled to your ears?

EXPLORE

Sound and Its Source

In the Explore Activity on pages 14–15, you saw that when you plucked or moved some common objects, they produced sounds. You noticed that each of the objects made different sounds. You also saw that the rubber band, the paper clip, and the ruler were moving back and forth rapidly in order to produce sounds.

Cymbals crashing together, clapping your hands, shouting, or closing your textbook are all actions that produce movements in the air.

*This rapid, back-and-forth movement produces sounds. This kind of movement is called **vibration** (vī brā´ shən).*

These are the same kinds of vibrations caused by beating a drum, plucking guitar strings, or playing the trumpet.

In each case, energy was transferred from something to the object that vibrated. Can you understand how sound is a form of energy? Try the activity to see how sound transfers energy.

Activity!

Sound Power

Here's an activity to help you understand how a vibrating object transfers energy to whatever it touches.

What You Need
overhead projector, plastic container of water, tuning fork, paper, *Activity Log* page 4

Place a plastic container half full of water on top of the overhead projector. Turn the projector on. Strike the tuning fork against the bottom of the heel of your shoe. (A tuning fork should never be struck against anything hard because it could be damaged.) Listen to the sound the tuning fork makes. Watch the ends of the tuning fork. Strike it again and quickly put the ends of the fork into the water. Watch the movement on the wall. Energy was transferred to the water. Was energy also transferred to the air?

Strike the tuning fork and hold it next to a piece of paper. What happened? Are you surprised that vibrating objects transfer energy to whatever they are in contact with?

You've seen how energy is transferred by the movement of a tuning fork. The fork did work as it transferred energy to the paper and the water.

At the same time, the tuning fork was also producing sound. It did this by transferring energy to the air.

Tuning fork

How Sound Travels

Energy is transferred by sound. Everything is made up of tiny atoms and molecules. Solid materials, water, and air are composed of molecules. When something vibrates it causes the molecules in the air to bump into each other. Because sound waves are caused and transmitted by vibrations of molecules, sound can only travel through matter. Sound travels better through some kinds of matter than it does through others. Try this activity to demonstrate how the waves of a spring toy are like sound waves.

The regular disturbances caused by these bumping molecules are called **waves**. Sound waves travel out in all directions from the source just like raindrops in a quiet pond. The raindrops are like a source of sound sending waves out in every direction.

TRY THIS

Activity!

See the Waves

You can't see sound waves, but you can see a model of how they move.

What You Need
spring toy, safety goggles, *Activity Log* page 5

With your goggles on, stretch the spring on the floor. Pinch a few coils of the spring together and suddenly let go of the coils. What happened? Move the end of the spring toward and away from your partner rapidly several times. What happened? Record your observations in your *Activity Log*.

Sound travels through water, one form of matter. Whales communicate by sound underwater.

In the activity with the spring toy, the coils of the spring represent the molecules in the air. Like the coils of the spring, the molecules in the air squeeze together and then move apart as the waves travel through them. You made one pulse in your spring, and then made a wave with several pulses. A single sound can contain hundreds of waves.

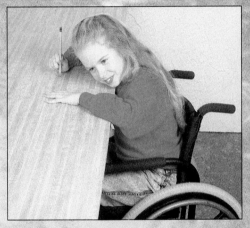

Sound travels through solids. For example, if you place your ear on a table, as you tap on it with your pencil you will hear the sounds that are produced. Do the next Try This Activity to see how else sound can travel.

Sound travels through air, too. When you speak, sound is produced and moves through the air.

TRY THIS Activity!

Spoon Hearing

Do sound waves travel only in air?

What You Need
metal spoon, string (60 cm), ruler, *Activity Log* page 6

Tie the handle of the spoon in the center of the string. Wrap the ends of the string around both index fingers, making sure both ends are about the same length. Place the ends of your index fingers in your ears. Lean over so the spoon hangs freely and tap it against the side of a table. How did the sound travel from the spoon to your ears?

Some Ways Vibrations Affect You

Groovy Vibrations

"Mary had a little lamb" were the first words ever recorded when Thomas Alva Edison made the first sound recording in 1877. Edison's early "phonographs" used waxed drums and were turned by a crank. Voice vibrations caused needle vibrations to be inscribed on the drum. When the drum was turned, it vibrated the needle, sending vibrations to the coneshaped speaker.

Today's record players still work much the same way. Grooves in the record cause a needle to vibrate as the record turns. Look at the grooves on an old record with a hand lens. In your **Activity Log** on page 7, sketch what you see in the record grooves.

Dangerous Vibrations

Jet airliners undergo dangerous stress from vibrations. Sound vibrations can cause small cracks to appear in the skin of a jet. Engineers search for these cracks by testing airplanes with the same thing that causes them—sound.

To test jet skin, air is forced from large blowers or air pumps into the jet being tested. This causes the plane to bulge slightly. Large amounts of air cause the cracks to open up a little bit, making a popping sound. When the cracks pop, they are detected by sound sensors. This method is also being used to find cracks in other objects that need to be kept airtight, such as railroad cars.

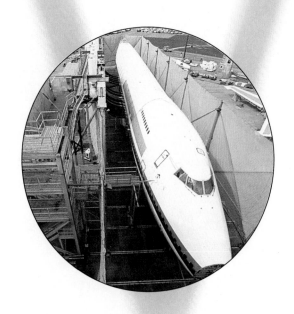

Feeling Vibrations

People who are deaf can dance even without hearing music. They can still sense the vibrations. They feel the vibrations carried through the floor and the air. They can move to the beat, or vibrations, of the music. See what vibrations you can feel. Try this activity.

TRY THIS

Feeling Sound

How can you feel sound?

What You Need
balloon, sound source with speaker, *Activity Log* page 8

With the sound source on, hold a balloon in front of the speaker. Do you feel the vibrations on the balloon? How are the vibrations transmitted? How else do you feel sound?

Sum It Up

Any rapidly-vibrating matter, even a paper clip, can make sound. Vibrations disturb the molecules in any gas, liquid, or solid that is touching the vibrating matter. This disturbance moves through matter as a sound wave. You hear sound because it transfers energy to your ears. This energy can have practical applications when it's used to warn or explain something. Loud sound can be harmful to your health. For the deaf it can be a valuable way of "feeling" the sound they can't hear. The energy of sound opens up a world that can be delightful or sometimes annoying.

Critical Thinking

1. What parts of your body vibrate when you talk?

2. If you couldn't hear any sound, how could you tell if a stereo speaker were working?

3. What makes the sound of a bumblebee's buzz?

People, machines, wind, and other objects make different sounds. Even a single object can produce various sounds. Why is it that outside on the playground you have to shout to be heard, but a whisper seems loud in an empty gymnasium?

WHAT DOES IT SOUND LIKE?

Minds On! Listen to some people talking. What differences are there between the voices of men and women? Do all girls have the same voices? Do all boys? How do people's voices change as they get older? Does your voice sound the same all the time? What are some conditions that cause your voice to sound different? Write your answers in *Activity Log* page 9. ●

You know that all drum sounds are caused by vibrations. But there must be more to it than that if even a simple drum can produce many different sounds. In the next activity, you will make a model that will tell you something about how one instrument can produce different sounds.

Activity!

Band Sounds

What makes sounds different? You'll experiment with a rubber band to see if it can produce more than one kind of sound.

What You Need

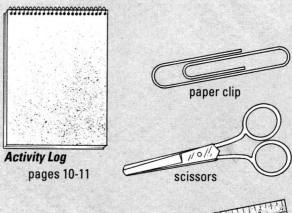

rubber band

tape

paper cup

Activity Log
pages 10-11

paper clip

scissors

ruler

What To Do

1 Punch a small hole in the bottom of the cup.

2 Put goggles on and cut the rubber band to make one long piece. Tie the rubber band to the paper clip. Now thread the rubber band through the bottom of the cup so the paper clip is held inside the cup.
Safety Tip: Use caution when using rubber bands.

Safety!

24

3 Tape the 30-cm end of the ruler to the side of the cup.

4 Predict what will happen when you pluck the rubber band. Hold the rubber band so it's stretched loosely and pluck it. What do you observe?

5 Stretch the rubber band over the end of the ruler and tape it down. Pluck the rubber band again.

6 Now pull the rubber band around the end of the ruler to make it tighter and pluck it again. Record your observations in your *Activity Log*.

7 Change the length of the part of the rubber band that vibrates by pressing your finger against the ruler at the 2-cm mark. Pluck the band again.

8 Now press your finger against the ruler at the 4-cm, 6-cm, and 8-cm points, plucking the band at each point. Record your observations in your *Activity Log*.

What Happened?

1. How were the vibrations different when you made the rubber band tighter?

2. How were the sounds different when you made the rubber band tighter?

3. How were the sounds different when you changed the length of the rubber band?

What Now?

1. How does changing the tightness of a rubber band affect the sound it makes?

2. How does changing the length of the rubber band affect the sound it makes?

3. What causes different types of sounds?

EXPLORE

Characteristics of Sound

By making a simple, one-stringed instrument, you were able to produce higher or lower sounds. The longer and the looser rubber bands made lower sounds. The shorter and the tighter rubber bands made higher sounds.

But what is different about the sound waves that produce higher or lower sounds?

The number of times an object vibrates in a given amount of time is called **frequency** *(frē´ kwən sē).*

If the string vibrates five times in one second, the frequency would be five vibrations per second. In a sound wave in air, the air molecules move back and forth as the wave passes, rather than up and down like the strings on the bass.

Hernando playing the bass.

TRY THIS
Activity! Zipper Pitches

Can a zipper have high and low pitches?

What You Need
zipper, *Activity Log* page 12

How can a zipper produce different pitches? Zip the zipper up and down, first very fast, then very slow. What do you hear? The zipper's teeth vibrate and cause the sound. Moving the zipper very slowly makes a low pitch, because the frequency is slower. What happens to the pitch when you move the zipper faster? Can you hear how the speed at which you move the zipper changes the sound?

Larger vibrations produce louder sounds. Smaller vibrations produce softer sounds. **Volume** is how loud or soft a sound is. If you pluck the rubber band on your model instrument harder, it will produce a louder sound. You're adding more energy to the string. More energy is applied to produce a louder sound. Think about how much effort it takes to shout as loudly as you can. How much effort does it take to talk softly?

When a sound has a high frequency, we say it has a high pitch. A sound with a low frequency has a low pitch. **Pitch** is how we perceive the frequency of sound. In the Explore Activity on pages 24–25, the loose rubber band vibrated slowly. It made a sound with a low frequency or low pitch. When you tightened the rubber band, it vibrated faster and made a higher-pitched sound.

Sound waves that you can hear have frequencies between 20 and 20,000 vibrations per second.

Focus on Environment

What Can You Do About Noise Pollution?

Did you know that the noise level of video-arcade games has been measured high enough to be considered hazardous? Even the roar inside your school bus or on the playground can be loud enough to cause ear damage. Power tools, music played at aerobic-dance classes, personal stereos, and loud music have caused increasing numbers of people to have hearing problems at an earlier age than ever before.

With your classmates, discuss what you can do to decrease noise pollution in your school. Then in your *Activity Log* on page 13, write a detailed plan of how to do it.

Speakers can make loud sounds.

Speed of Sound

Which comes first in an electrical storm—the flash of lightning or the sound of thunder? Even when the storm is far away you see the flash before you hear the crash. Sound waves travel much more slowly than light waves. Sound travels about 340 meters (1,115 feet) in one second. Light travels about 300 million meters (186 thousand miles) in one second. Because light travels so fast, we see things almost instantly.

Because sound travels more slowly than light, we sometimes see things before we hear them. If Kelly were standing 340 meters away from Regina and Kelly shouted, Regina would see Kelly's lips move one second before Regina heard the sound. You can use this difference to estimate distances.

Kelly

Math Link

When Was It Heard?

You can use the difference in the speed of sound and light to measure distances. Suppose you are watching a baseball batter. The time from when you see the batter hit the ball until you hear the crack of the bat is 1.5 seconds. The distance to the batter equals the speed of sound multiplied by the time it takes to hear the sound. The speed of sound is 340 meters (1,115 feet) per second. The time is 1.5 seconds. How many meters are you from home plate? Write the problem in your **Activity Log** on page 14. Use a calculator to get your answer?

How long before you hear the sound of the bat hitting the ball?

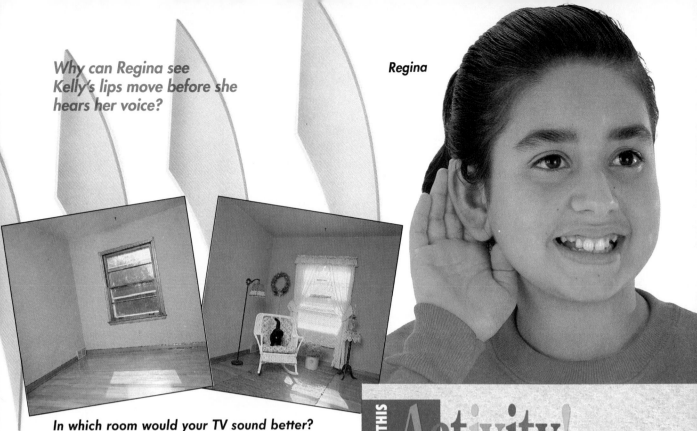

Why can Regina see Kelly's lips move before she hears her voice?

Regina

In which room would your TV sound better?

Some materials do not bounce sound waves well. The sound waves are **absorbed** (ab sôrbd´) or taken in by the surfaces in the furnished room.

Once sound waves are set in motion, some of them are going to be absorbed by objects. Others will bounce off objects. Smooth, hard, flat surfaces bounce sound. Soft, irregular surfaces absorb sound. The carpet and furniture absorb some of the sound. Hard things, like wood floors or walls, reflect more sound waves than they absorb.

Why do you think you hear a lot of noise when you move through an empty house, but hear very little noise after carpets and furniture have been moved in?

TRY THIS

Activity!

Sound Absorption

How can you soundproof a coffee can?

What You Need
various materials, coffee can, *Activity Log* page 14

Yelling into a coffee can makes a loud sound. What kinds of things can you put in the can to absorb the sound? Yell into the empty can. How loud is it? Now put some materials in the can, like the cotton, paper, foil, etc. Now yell into the can with all the materials in it. How does adding the materials to the can affect the loudness of the sound? Why do you think the loudness of the sound changed after you added the materials?

Bouncing Sources

The next time you go to a theater or auditorium, look at the decorative wall construction. Theater designers use the kind of information you have been studying about sound.

Activity!

How Can You Make It Louder?

You just learned how sound can be absorbed. Now you'll learn how sound bounces off objects.

What You Need

cardboard tube, *Activity Log* page 15

With your mouth close to the tube, shout into the tube and listen to the sound. Now shout into the tube holding your hand over the end of the tube. Try to shout with the same force. Notice how loud you sound this time. In your *Activity Log,* draw two pictures of the times you shouted into the tube and indicate the sound waves. Why does it sound louder when you shout into the tube with your hand on the bottom?

The second time you shouted into the tube it sounded louder because the sound bounced back towards you. The next time you shouted the sound traveled out. Therefore, it wasn't as loud.

The quality of sound in the room can be controlled by the shape of the room and the materials used to absorb or reflect sound.

An **echo** (ek´ō) is the repeated sound made when sound waves are bounced off of a surface.

Gymnasium

Amphitheater

If you shout in an empty gymnasium, your words bounce right back at you. How quickly you hear the echo depends upon how far away you are from the surface that's bouncing the sound.

Hello ----- H-e-l-l-o

Minds On!
Have you ever shouted across a valley in the mountains? If you have, you probably noticed that your voice echoed. Why do you think this happened? ●

The High Frequencies of Ultrasound

People use echoes to measure distances in the ocean. A **sonar** (<u>So</u>und <u>Na</u>vigation and <u>R</u>anging) device sends out sound waves from ships and measures the time it takes for the waves to bounce off an object and return. Treasure hunters use sonar to locate sunken shipwrecks. The sound waves used in sonar are called ultrasonic. **Ultrasonic** (ul´ trə son´ ik) sounds, or ultrasound, have frequencies that are too high for the human ear to hear.

Some animals have their own sonar. Bats make ultrasonic noises and use their big ears to catch the echoing sound waves. The echoes help the bat catch food from the air and avoid objects while flying in the darkness.

Dolphins use echoes to find or avoid objects in the ocean. They are so good at translating echoes that they can tell the difference between a nickel and a dime dropped into the water.

Fishing boats use sonar to find fish.

32

SCIENCE TECHNOLOGY AND Society

Focus on Technology

Seeing With Sound

Even though ultrasound frequency is too high for humans to hear, we still make good use of it. These high-frequency sounds are used in hospitals to examine the human body. The bouncing echoes of the sound signals are used to make images of internal body organs in much the same way that sonar technology is used to locate objects in water.

One especially useful way doctors use ultrasound is to see unborn babies. An instrument called the ultrasound scanner can show doctors how the unborn baby looks inside the mother.

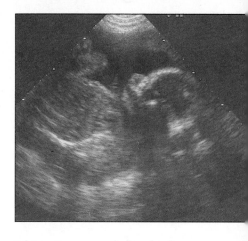

This picture of the baby allows doctors to make sure everything is going okay with the baby before it's even born!

Sum It Up

How is energy used to make sounds? The one-stringed instrument you made was making sound because you applied energy that interacted with the rubber band on the ruler. The frequency of vibrations was higher or lower depending upon the length or tightness of the band. The pitch of the sound depended upon the frequency. You could control the volume of the sound by how much energy you used to pluck the band.

Sound waves are sometimes absorbed by objects and other times bounced off of objects, creating an echo. Whether they are absorbed or bounced back, you know that all sounds occur because of energy interacting with matter.

Critical Thinking

1. You can make your voice high-pitched or low-pitched. What do you think changes in your body to make the pitch change? How could you find out?

2. When string on a guitar or other stringed instrument is tightened, what is the result of the pitch? Why?

3. How is the volume changed on an instrument?

Day after day, the sun produces enough energy to send light out to all the planets in the solar system, including Earth. But the moon produces no light at all. So how does the moon light our way through the darkness of night? Where does most of the light on Earth come from and what causes that light?

WHAT IS LIGHT?

Long ago, people discovered that fire could produce light. Next came torches, candles, oil lamps, and gas and electric lights. We even use batteries to store energy to make light in flashlights. Natural or artificial, light may be a mystery to you.

You may think you don't know much about light, but you do. Light has some properties similar to sound. As we study light, watch for similarities and differences between light and sound.

Minds On!
Mirrors tell us a lot about light. You may use mirrors mostly for grooming. In your *Activity Log* on page 16 list some other ways mirrors are used. ●

Many of the uses we make of light are possible only because we know something about how light travels from one point to another. Mirrors tell us something very important about how light travels. You will have the chance to experiment with traveling light in the next activity.

A day at the beach

Activity!

Bouncing Light

How does light travel? Does it travel in a straight line? Can it bend around corners? What happens when it hits something? The answers to these questions are very important in our study of light, as you are about to see. Using mirrors, you and your partner will complete 3 different tasks in this activity.

What You Need

Activity Log pages 17-18

textbook

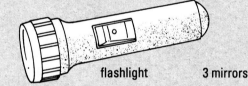

flashlight

3 mirrors

What To Do

1 Lean 1 of your mirrors against the textbook so it will stand up.

2 Write your name on a piece of paper and look at it in the mirror. What did the mirror do to your name?

3 Look in your mirror. Touch the right side of your face. Which side of your face did the mirror image touch?

4 Stand the textbook up in the middle of your desk. Point the flashlight at the front side of the textbook and turn it on.

5 Without moving the textbook or the flashlight, cooperate with your partner and use the 3 mirrors to make the flashlight beam shine on the backside of the textbook. Try doing this many different ways. Then, draw diagrams in your *Activity Log* of the paths you made the light take to get to the other side of the textbook. Show where the mirrors were.

What Happened?

1. How were the images different in the mirror?
2. How many ways did you bounce the light off mirrors to make it go to the other side of the textbook?

What Now?

1. Draw a diagram in your *Activity Log* of how you and your partner would place your 3 mirrors to make the flashlight shine into the hallway from your desk.
2. What are some ways you could make use of bouncing light?
3. How does light travel around corners? What path does it take?

EXPLORE

Characteristics of Light

When you looked in the mirror in the Explore Activity, you saw a backward image of yourself. That image tells you something important about light. An **image** is any picture of an object made by a mirror or lens.

Smooth, flat objects such as shiny pans and foil paper bounce or **reflect** light better than rough objects. When light hits a smooth surface, most of the light bounces off. Light travels in a straight line. When you used the mirrors in the Explore Activity on pages 36–37, you observed how light rays travel in a straight line. When light bounces off a rough surface, it is scattered in many directions. How is this like the sound waves you studied in the last lesson?

All objects reflect light, but some, particularly mirrors, do a better job of it than others.

38

The reason you see your image in the mirror is because mirrors reflect light.

Being able to see the reflection in the mirror of what's behind her helps the bicycler stay safer when she's riding.

It's fortunate for us that light bounces off objects, since reflected light is what enables us to see things. If you could see only those things that emit their own light, your world would be a very dark place. Very few objects send out their own light. Can you name some?

To see examples of objects that emit their own light and objects seen only by reflected light, all you have to do is look toward the sky. The sun and stars are visible because they send out their own light. The moon and the planets do not have their own light and are visible only because of the sun's light reflecting off them.

Bioluminescent fireflies

Bioluminescent squid

If you've ever caught fireflies, you know that some animals can make their own light. Shrimp, squid, and some fish are also able to produce light in special body cells. This ability of living things to produce light is called **bioluminescence** (bī ō lü mə nes´ əns). The energy to make the molecules vibrate comes from chemical reactions that give off light, not heat.

Just like sound travels through solids, liquids, and gases, so does light. But as light travels, it is blocked by many materials.

Light energy can also be absorbed. When white light strikes a colored object, some colors of light energy are absorbed. The object will then transmit or reflect the colors of light that have not been absorbed, which is why we only see certain colors.

Things that prevent light from passing are called **opaque** (ō pāk´). Wood, iron, and most of your body parts are opaque. They reflect or absorb light energy.

Matter that allows light to pass through it clearly is **transparent** (trans pâr´ənt). Water, air and some glass are transparent. They transmit the light energy and do not absorb it.

Objects that allow some light to pass through but scatter some of the light are called **translucent** (trans lū´ sənt). Waxed paper, paper, and some plastics are translucent.

Opaque

Transparent

Translucent

TRY THIS Activity!

Appearing Coin

What happens to light when it hits water? Here's an activity that illustrates how the path of light changes.

What You Need
coin, water, opaque cup, *Activity Log* **page 19**

Place a coin in the opaque cup as shown in the picture. Stand back until the coin is hidden from your sight. Now have a classmate slowly pour water into the cup. What happens?

The reason you can see the coin after the water is poured in the cup is because the light from the coin changes direction as it travels to your eyes. The light is bent as it passes from the water to the air.

Bending Light

Light waves normally travel in straight lines. When those light waves move from one kind of matter to another kind, like from air to water, their speed changes. And the change in speed causes the light rays to bend. The bending of light is called **refraction** (ri frak´ shən). The light in the cup with the coin was refracted because the light traveled through a different material and changed speed.

Imagine a group of people running in a line, side-by-side.

When the people on the right side of the line suddenly hit the water, they slow down.

When everyone is in the water and moving at the same speed again, the line would be straight but moving at a different angle.

41

Lenses

Refraction is what makes lenses work. There are two basic types of lenses.

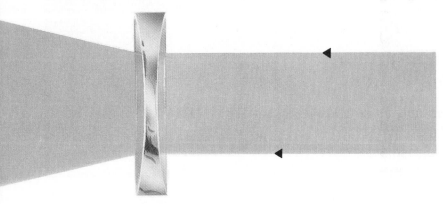

Concave lenses are thinner in the middle than they are on the edges. Concave lenses make objects look smaller than they really are.

In concave lenses the light rays hit the lens and pass through. The light rays spread out in all directions.

Convex lenses are thicker in the middle than they are on the edges. Convex lenses make objects look bigger than they really are. Magnifying glasses are really convex lenses.

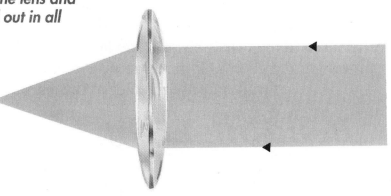

In convex lenses the light rays hit the lens and pass through. The light rays move into a point.

Literature Link
Rainbows to Lasers

Use the index to look up the word *lens* in the book *Rainbows to Lasers* by Kathryn Whyman. Read about lenses. In your **Activity Log** on page 20, write down the new things you learned about lenses. Now, look at the section in your book about eye problems. It explains that some people need glasses with different types of lenses to see near and far.

Draw a picture in your **Activity Log** that shows what kind of lenses are used for people who can't see far or can't see near. Take a poll of your classmates to see how many of them wear glasses. Look to see if the lenses of their glasses are concave or convex. Compare your conclusions with four of your classmates.

Telescope

Camera

Microscope

Lenses are inside telescopes, cameras, and microscopes. What other things have lenses in them? Water can also be used as a lens. A water lens is very similar to a magnifying glass.

TRY THIS

Activity!

Water Lens

Have you ever noticed how a drop of water sits on a flat surface like a rounded bubble? What kind of lens would a water drop be like, convex or concave?

What You Need
foam cup, plastic wrap, transparent tape, dropper, objects, newspaper, water, scissors, *Activity Log* **page 21**

Carefully cut the half-inch rim from the foam cup. Then, from a piece of plastic wrap, cut a circle about an inch larger than the rim of the cup. Stretch the plastic tightly over the rim. Tape it down and under the edge. Place a few drops of water on top of each other in the center of the plastic using a dropper. This is your water lens.

Now place objects and newspaper under the water lens and notice the size of the objects. Why do you suppose the objects and the newspaper looked the way they did through the water lens? What happens to the image as you add more or less water? Try it and see. Draw the image in your *Activity Log*. Why do you think a water lens is similar to a magnifying glass?

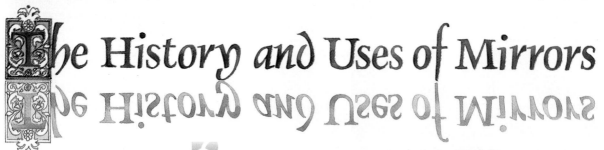

The History and Uses of Mirrors

Literature Link
Fascinating Reflections

People's fascination with their reflected images started with the first person who looked into a pool of water. In fact, a character in an ancient Greek myth named Narcissus (när sis´ əs) ran into some problems because of this.

According to the myth, Narcissus went to a fountain one day to get a drink. When he saw his reflection in the water, " . . . he thought it was some beautiful water-spirit living in the fountain. He stood gazing with admiration. . . [and] fell in love with himself" *(Bulfinch's Mythology).*

After that, Narcissus spent the rest of his life staring at his reflection. He died without ever falling in love with another human being.

Find the words *narcissistic* and *narcissism* in a dictionary. Explain what each word means. What else does the word *narcissus* refer to besides the mythical character?

Many people besides Narcissus have enjoyed looking at their own reflections since early times. But objects similar to the mirrors we use today didn't begin to appear until about 3500 B.C., when the Sumerians in Mesopotamia (mes´ ə pə tā´ mē ə) began using polished metal to make reflecting devices.

Silver and gold were popular materials for early reflecting surfaces, and so valued that servants were sometimes given mirrors as part of their pay.

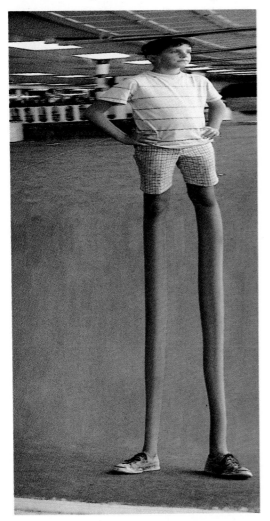

It wasn't until 1840 that mirrors began to be produced by coating glass with a backing of silver. Today, both silver and aluminum are used on the back surface of ordinary glass mirrors.

Much later, glass mirrors were used in Italy in 1300. Decorative wall mirrors began to be popular as early as the 15th century.

Mirrors are made out of thin sheets of aluminum or silver that are put onto glass. The light hits the glass and then is reflected off.

There are three basic types of mirrors.

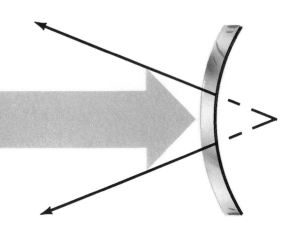

The one you look in every morning is a flat mirror. In the Explore Activity on pages 36–37, you used flat mirrors to reflect the light of a flashlight.

The second type of mirror is called convex. The passenger side mirror of a car is convex. The shape of a convex mirror is curved outward so the images produced will be smaller than the object. This mirror in a car produces images smaller than the object. This allows the driver to see a wider angle of the road. What other types of convex mirrors have you seen?

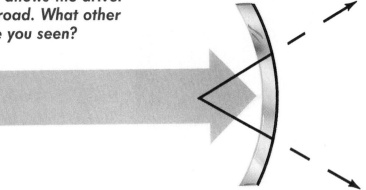

The third type of mirror is called concave. Makeup and shaving mirrors are concave. The shape of a concave mirror is curved inward, like a bowl. A concave mirror magnifies the object's image. Flashlights, spotlights and headlights are made with concave mirrors.

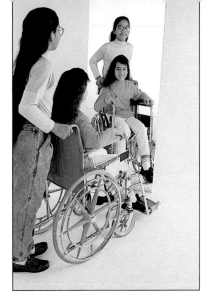
Flat mirror

Convex mirror

Concave mirror

Mirrors should not be confused with lenses. There are convex and concave lenses also. Remember that lenses are glass through which light can pass. Mirrors reflect light.

Minds On! In your ***Activity Log*** on page 22, make a chart comparing lenses and mirrors. How are lenses and mirrors different and alike? Which do you use more often in a day? ●

Sum It Up_____

Like sound, light transfers energy in waves. Light can pass through empty space, liquids, gases, and even some solids. Without light to reflect off objects, we couldn't see most things. Objects react differently to light. For example, opaque objects absorb or reflect light so it can't pass through. Transparent objects allow light to pass through clearly. Translucent objects allow some light through.

Although light energy travels in straight lines, we've learned to redirect its path with mirrors and lenses. Mirrors and lenses are part of things you use to see with, like eyeglasses, binoculars, cameras, microscopes, and telescopes.

Critical Thinking

1. Why are some mirrors and lenses curved?

2. If you were designing an underground house, how would you use mirrors to get sunlight into all the rooms?

3. How big does a mirror have to be to see all of yourself in it?

The sun, we say, is yellow. But sometimes it looks red. It sends white light across a sky that looks blue and shines on grass that looks green. In the shade, the grass looks dark green. Is color really that confusing? In this lesson, you will explore how light interacts with the objects around us.

WHAT COLOR IS IT?

Colors are all around us—millions of shades of them. Where do these colors come from? Why do they sometimes change before your very eyes?

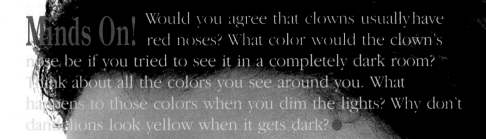

Minds On! Would you agree that clowns usually have red noses? What color would the clown's nose be if you tried to see it in a completely dark room? Think about all the colors you see around you. What happens to those colors when you dim the lights? Why don't dandelions look yellow when it gets dark? •

A world without the brightness and variety of colors would be like living in a black-and-white television movie. Red stop signs and lights warn us of danger. We cheer for athletic teams that we know by the color of their uniforms. Nature is filled with brightly colored plants and animals. People plant colorful flowers to brighten up their yards.

The upcoming activity will give you a chance to shine some light on the mystery of where all those bright colors come from.

Activity!

How To Get White Light

In this activity, you and your partners will experiment to see how white light is produced.

What You Need

4 pieces of construction paper (white, red, blue, green), 30-cm square of each

Activity Log pages 23-24

3 flashlights

transparent tape

3 pieces of cellophane (red, blue, green), large enough to fit over the head of a flashlight

What To Do

1 Tape a piece of cellophane over the bulb end of each flashlight.

2 Turn out the lights. Cooperate with your partners to shine the 3 flashlights on the white piece of construction paper so all the light beams meet.

3 Experiment with the flashlights, moving the beams around so they cross over each other until you get white light.

4 Record your results in your *Activity Log*.

5 Now, shine the flashlights on the different-colored pieces of construction paper. Look at the paper through the different-colored filters.

6 Record your observations in your *Activity Log*.

What Happened?

1. What color light beams did you combine to produce a white light spot?
2. What other color light spots did you get?
3. What happened when you looked at the papers through the different-colored filters?

What Now?

1. What color would a red rose appear if it were lit by only blue light? What color would it appear if lit by only green light?
2. Why weren't different-colored beams reflected the same from different-colors of paper?

EXPLORE

White Light

We normally think of sunlight and the light from a flashlight or light bulbs as being white, or having no color at all. But as you saw in your activity, white light does contain colors. It's pretty amazing, isn't it?

If you've ever seen a rainbow in the sky, then you have seen an example of how colors come from white light.

Theater lighting designers must be able to blend light inside a theater so scenes look like natural settings.

Because white light is a combination of all colors, the colored beams shining on the student mix to form a white spot.

Theater Lighting Designer

Theater lighting designers have to know about color and light. They need to know how different colors of costumes will look in different colors of light.

Students in college or university theater programs are trained to

become lighting designers. Lighting design is also important in television and film. There aren't many jobs available, but it can be fun for those who enjoy being part of the behind-the-scenes crew of a play, film, or TV show.

TRY THIS Activity!

Spinning Colors

What happens when colored objects move very fast? Will you still be able to see all the separate colors? Find out!

What You Need
10-cm heavy cardboard disk, rubber band, colored pencils (red, orange, yellow, green, blue, and violet), protractor, *Activity Log* page 25

Using your protractor, divide the cardboard disk into 6 equal sections of 60° each. Color the sections in the disk in the order the colors are listed. Now make 2 small holes in the middle of the disk, about 1 cm apart. Thread the rubber band through the holes and tie it together. Pull the rubber band tight. Twist the rubber band up as shown. Pull the rubber band tight. What do you see as the disk spins?

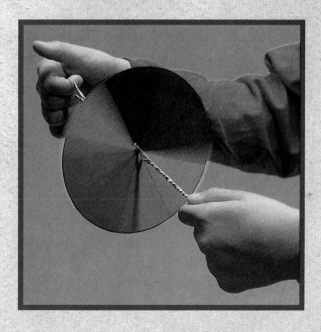

When you spun the disk quickly, your eyes could not see each reflected color separately. You saw only the result of mixing the different colors together. Remember that white light is a mixture of all the colors of visible light.

X rays

Ultraviolet light

Visible light

Infared waves

Microwaves

Radio

Visible Light

The colors you are able to see are called **visible light**. And visible light is part of something we call the **electromagnetic spectrum** (i lek´ trō mag net´ ik spek´ trəm). It's made up of electrical and magnetic energy. The **spectrum** is made up of visible light and other kinds of waves that we can't see, such as radio waves, microwaves, and X rays.

The colors of the spectrum you see are always in the same order—red, orange, yellow, green, blue, indigo (deep blue), and violet. To remember the order of colors, think of the name made by their initials, ROY G. BIV.

White light

*A miniature rainbow is formed by a **prism**. A prism (priz´ əm) is a piece of glass that's specially cut and polished, like crystals you might see people hang in their windows to catch the sunlight.*

When light is shined through the prism, it separates the white light into all the colors of the rainbow.

When you see a rainbow, it's because tiny drops of water in the air act just like prisms to refract sunlight.

You see colored objects the way you do because of light absorption and light reflection.

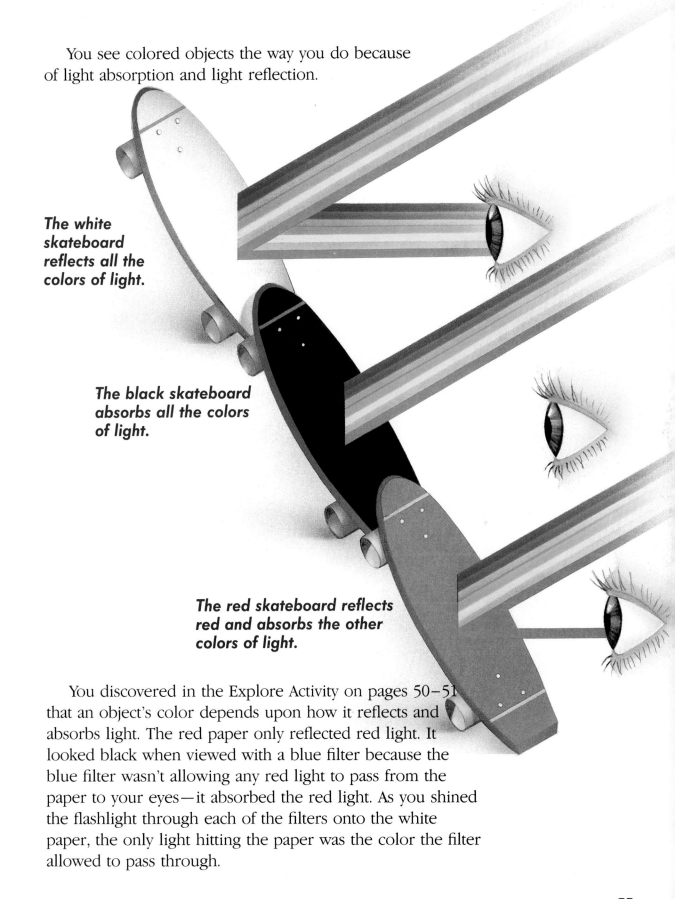

The white skateboard reflects all the colors of light.

The black skateboard absorbs all the colors of light.

The red skateboard reflects red and absorbs the other colors of light.

You discovered in the Explore Activity on pages 50–51 that an object's color depends upon how it reflects and absorbs light. The red paper only reflected red light. It looked black when viewed with a blue filter because the blue filter wasn't allowing any red light to pass from the paper to your eyes—it absorbed the red light. As you shined the flashlight through each of the filters onto the white paper, the only light hitting the paper was the color the filter allowed to pass through.

Mixing Pigments

You know that a piece of red paper appears red because it reflects only the red color and absorbs all other colors of light. But you may be wondering what causes it to reflect the red color in the first place. An object takes its color from the matter it is made of or the coloring material that has been applied to it.

Pigments (pig´ mənts) are solid particles that produce colors by absorbing or reflecting light.

Pigments are used to make the colors of paint these students are using in the class.

Whether an object is colored by natural pigments or by pigments mixed into paint, the color is seen because of the way light reflects off matter. The student's blue jeans are blue, and his apron is green—all because of the way they reflect light waves to your eyes.

It may seem strange, but colored lights mix differently than colored paint pigments. Red, green and blue lights are mixed together to produce other colors of light. When colored lights are mixed, lights of different colors are added together to create new colors. But an artist mixing paint uses red, yellow, and blue. When pigments in the red, yellow and blue paint are mixed, new colors result from colors being subtracted. Sounds complicated, doesn't it? Here's an activity to help you see for yourself how this works.

The colors that make up the pictures in this book, for example, are determined by substances in the ink called pigments.

TRY THIS Activity! A Magic Mix

What do you think happens when paints are mixed?

What You Need
3 colors of paint—red, blue, and yellow—plastic spoon, pan, *Activity Log* page 26

Before you start, think about what colors the paints are reflecting and absorbing. Now, put a little of each of the paints in the pan and mix them together with the spoon. What color did you make? What colors of light did you subtract when you mixed all the colors together? Record your observations in your *Activity Log*.

Remember that red paint reflects red and absorbs all other colors. Blue reflects blue and absorbs all other colors. Yellow reflects yellow and absorbs all other colors. When you mix them together, every color of light is absorbed. What you have left is black, because all the light is absorbed and none of it is reflected.

Dots of Color

**George Seurat applied paint one dot at a time.
His style was called _pointillism_ (poin´ til izm).**

Did you know that the image on your TV screen is made up of many horizontal lines of tiny colored dots? These dots are like the picture you just looked at in the magazine, and like Seurat's painting style.

When you paint or color a picture, you probably make a sweeping stroke with the brush or crayon. But there is another way to use colors to form a picture.

For example, George Seurat's paintings are made up of thousands of tiny, colored dots, side-by-side. When you stand back and look at the picture from a distance, the dots blend together. Seurat developed his approach after studying the new theories of light and color which were just becoming popular at the time. The colors look brighter because your eye blends them, not a brush on canvas.

Look at a colored picture from a magazine with a hand lens. Do you notice anything? What colors can you distinguish?

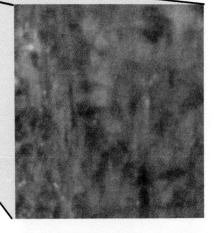

HIGH RESOLUTION TV

By the year 2000, an improved type of color TV called high resolution television may make today's televisions seem out-of-date. High resolution television produces a picture sharp enough to be blown up to wall size. The picture is so good you might think you're sitting in a theater. What makes it so good? Like regular television, high resolution screens are made up of many horizontal lines of tiny colored dots, but high resolution TV sets have twice as many lines of dots as regular televisions.

Tiny dots make up high resolution TV pictures.

Sum It Up

The next time you turn on a light, you'll know you have all the colors of the rainbow in front of you. But you won't be able to see all the colors at once. Which ones you see depends entirely upon which waves are absorbed by and which are reflected from the objects around you. The visible light is the only part of the electromagnetic spectrum you can see.

An object is the color it is because it only reflects its own color. The other colors are absorbed. White light is white because white objects reflect all colors. Black is black because all the colors are absorbed. Pigments are what make objects different colors. Different pigments reflect different combinations of colors. We use color in many ways to add pleasure and interest to our lives.

Critical Thinking

1. What colors would you see if you wanted to absorb all the energy of light striking an object? What if you didn't want any light energy to be absorbed?

2. What colors absorb more sunlight than others? Why?

3. Some animals use color to find food or each other. Can you think of some of these animals?

The musician blows into his tuba. The instrument's molecules bump against air molecules and send out sound waves in your direction. The sunlight reflects off the band's brightly-colored uniforms. All those sound and light waves would pass you right by if you didn't have a special set of receiving equipment—your eyes and your ears.

HOW DO WE SEE AND HEAR?

Don't look at the bright sun. Don't turn your stereo up so loud. Quit reading in the dark. Don't stick anything in your ear. Have you ever been warned about things that can damage your ears or eyes? There's a good reason. Seeing and hearing tell you almost everything you know about what's going on around you.

Your eyes and your ears allow you to enjoy the sights and sounds of a parade.

Minds On! Why are your ears shaped the way they are? Why is there one on each side of your head? Would you be better off with bigger ears? Sit very quietly and listen to the sounds in the room. Now cup your hands around both ears. What difference do you notice about sounds? ●

Our eyes and ears are sense organs that help us capture the energy in sound and light waves around us. Without them, those messages would go unnoticed.

How do your ears and eyes turn the energy of sound vibrations and light waves into something you can use? In the next activity, you'll make a model that will tell you something about how the parts of the ear transmit different sounds to your brain.

61

Activity!

How Do We Hear?

How do your ears turn all those vibrations in the air into something that makes sense? Here's an easy activity that will give you a better idea of the kinds of parts that make up your ear and how they react to sound waves.

What You Need

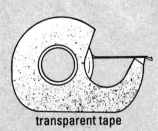
transparent tape

Activity Log pages 27-28

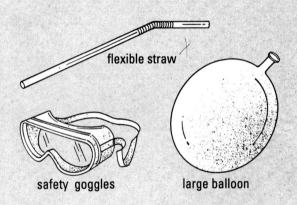
flexible straw

safety goggles

large balloon

cup

water

coffee can

What To Do

1 Cut the neck off the balloon first, then stretch the balloon over one end of the coffee can.
Safety Tip: Watch jagged edges on the coffee can.

Safety!

62   See the *Safety Tips* in steps 1 and 2.

2 Put goggles on and tape the straw to the center of the balloon so it sticks out to the side. *Safety Tip:* Be sure the balloon is secured to the coffee can.

3 Lay the can on its side and tape it to your desk.

4 Bend the short end of the straw so it points down and put it into the glass of water.

5 Have your partner watch the glass of water while you make various noises by the open end of the coffee can. Record your observations in your *Activity Log*.

What Happened?

1. What moved when you made noises into the coffee can?

2. Did making louder or softer sounds make a difference?

What Now?

1. Why did the straw move?

2. If you couldn't hear any noise, how could you tell that your partner was making noise near the coffee can?

EXPLORE

The Ear

The coffee-can experiment is a good illustration of sound energy being transferred to another object—in this case, the water in the glass. This simple model of an ear shows how sound waves are transferred to water. In a similar manner, sound waves interact with the structure of our ears to give us sounds.

The parts of our ears we see are the funnel-shaped flaps on the sides of our heads. But these outer ears are only part of the structure that turns vibrating waves into what we call sound.

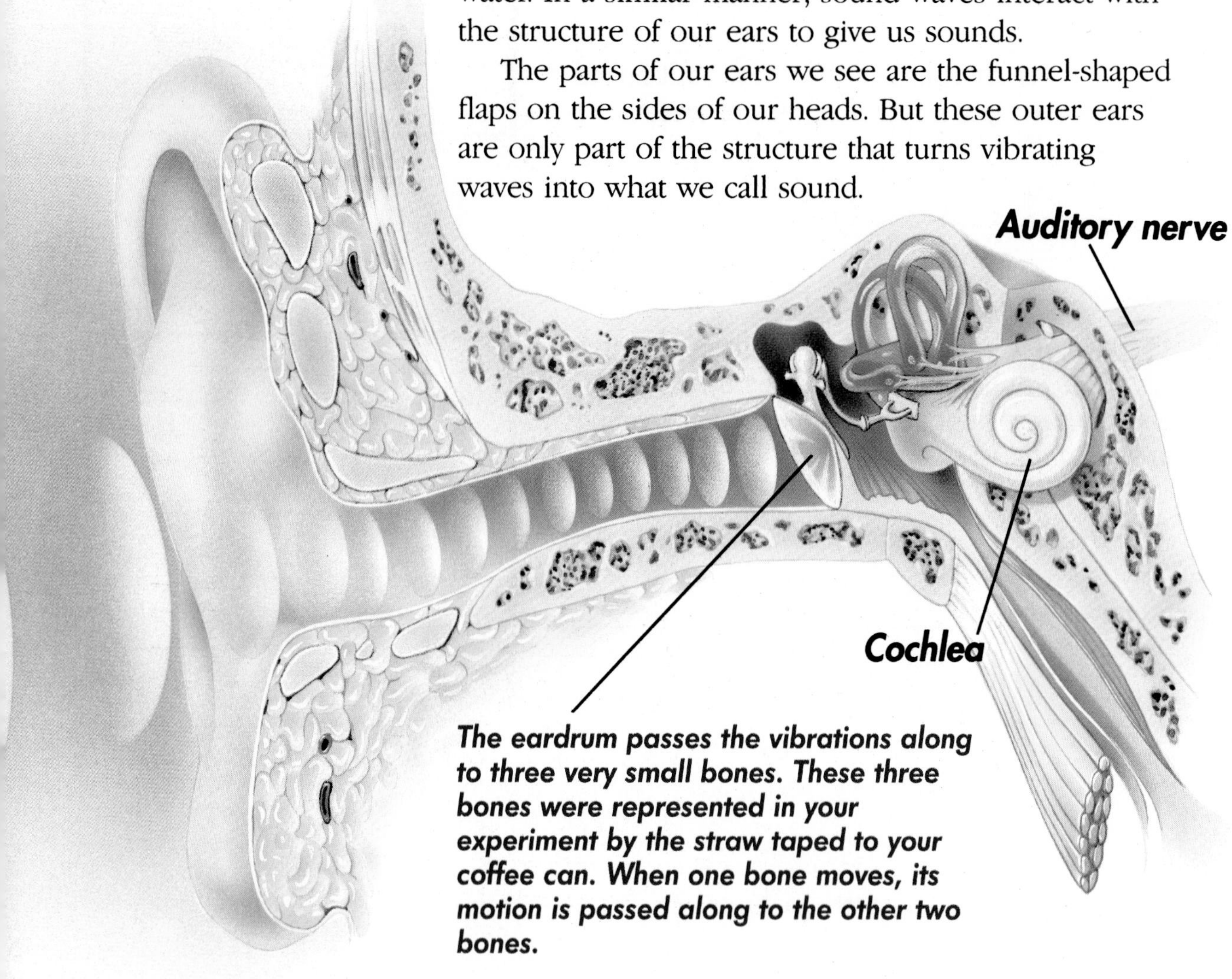

Auditory nerve

Cochlea

The eardrum passes the vibrations along to three very small bones. These three bones were represented in your experiment by the straw taped to your coffee can. When one bone moves, its motion is passed along to the other two bones.

When you hear noise, it's because sound is collected by the outer ear and funneled into the eardrum. The eardrum is like the balloon on your coffee-can model. Sound waves make the eardrum vibrate.

Telephones allow you to communicate with others.

From these three bones, the sound is passed on once again to an organ shaped like a coiled seashell called the **cochlea** (kok´lēə). *The glass of water represents the cochlea in your coffee-can model. The cochlea is filled with fluid.*

Having a hearing test

In the cochlea, the vibrations are changed into electrical messages that are sent to your brain through the **auditory nerve** (ô´ di tôr ē nûrv). *The brain interprets the message it receives and we recognize it as a particular sound.*

Catching sound waves is only part of our job if we want to know all we can about what's going on in our surroundings. There are many light waves that carry light in our direction. To catch them, our eye muscles are constantly focusing and refocusing, as many as 100 thousand times a day.

65

The Eye

Light enters your eye through an opening called the **pupil**, the black spot you see in the middle of your eye when you look in a mirror. In bright light, the pupil becomes smaller, allowing less light to enter the eye. This small hole protects the eye from bright lights and improves the image that you see.

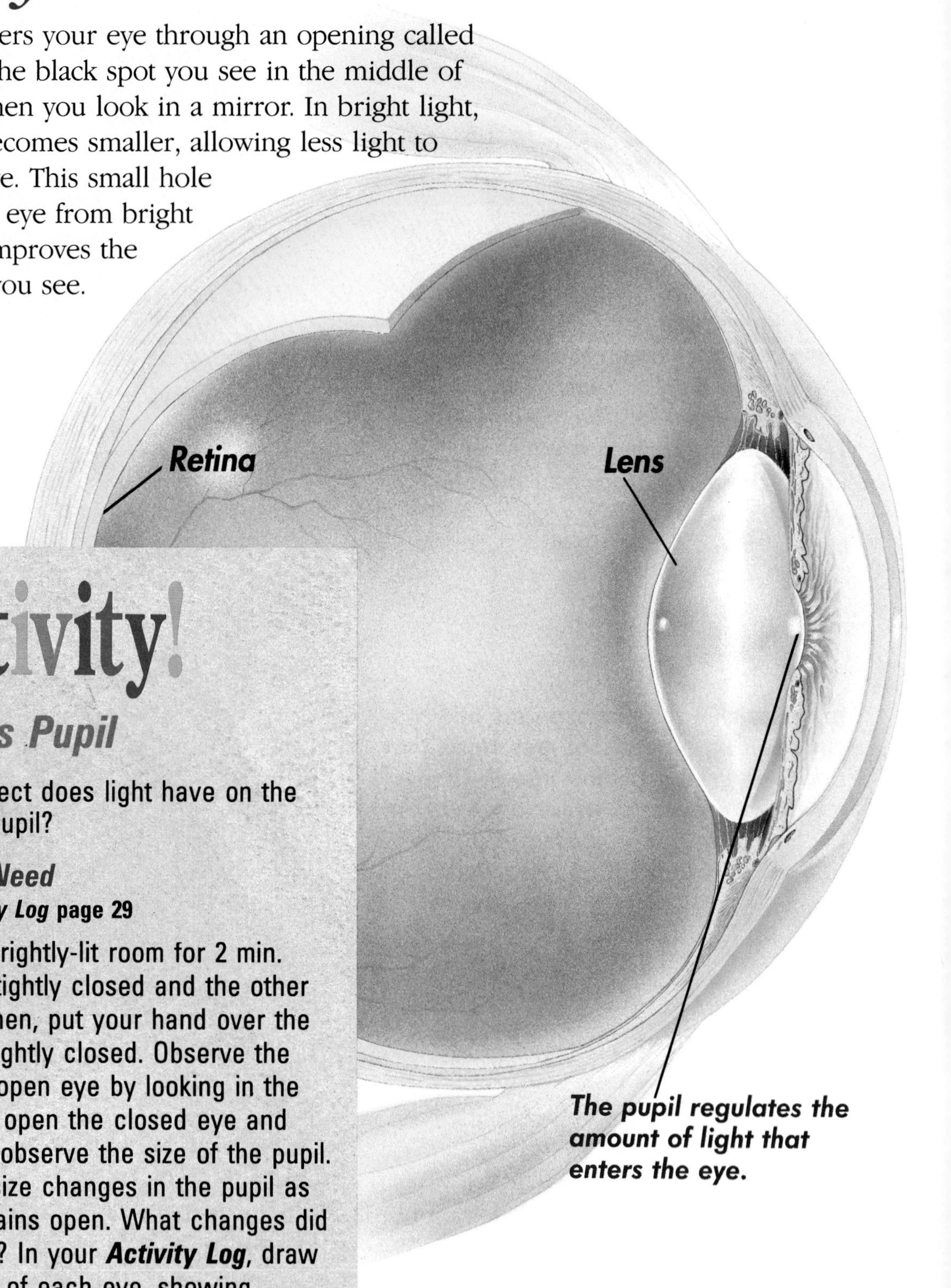

Retina

Lens

TRY THIS

Activity!

A Pupil's Pupil

What effect does light have on the size of the pupil?

What You Need
mirror, *Activity Log* page 29

Sit in a brightly-lit room for 2 min. Keep 1 eye tightly closed and the other eye open. Then, put your hand over the eye that is tightly closed. Observe the pupil of the open eye by looking in the mirror. Now, open the closed eye and immediately observe the size of the pupil. Notice any size changes in the pupil as the eye remains open. What changes did you observe? In your *Activity Log*, draw 2 pictures, 1 of each eye, showing the pupils.

The pupil regulates the amount of light that enters the eye.

The **lens** then focuses the light on the **retina** (ret´ə nə) at the back of your eyeball.

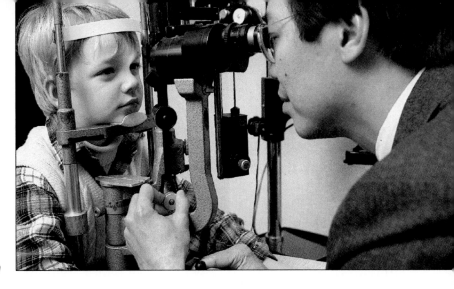
Eye exam

If things stopped with the pupil, we would have a problem. The image focused on the retina is upside-down. A message is sent along a nerve to your brain and the brain knows how to turn that upside-down image right-side up. Try this activity to see how the image looks before the brain does this.

TRY THIS Activity! *Upside-Down Is Okay*

Here's a way to see how your eyes work.

What You Need
flashlight, tissue paper, masking tape, black marker, hand lens, *Activity Log* page 30

On the center of the tissue paper, draw a thick arrow. Tape the tissue paper with the arrow in the center to the head of the flashlight. Shine the flashlight with the arrow pointing up on a wall. Hold the lens between the light and the wall. Slowly move the flashlight until you see a clear image of the arrow through the lens on the wall. What did you see? The image you saw was upside-down—just like the image that forms on the back of your eye.

Languages of Communication

Sound and light are so much a part of our lives. But some people are without the senses of sight and hearing. However, with the help of computers and electronic devices that can code sounds and words, they are able to communicate in the world.

Sign language is a way for the deaf and hearing impaired to communicate using their hands. This is also a way for people with normal hearing to communicate with the deaf and hearing impaired. Sign language consists of signs for many words and an alphabet to "fingerspell" others. Look at the hand signals for the sign language alphabet and see how you would fingerspell your name. Try it.

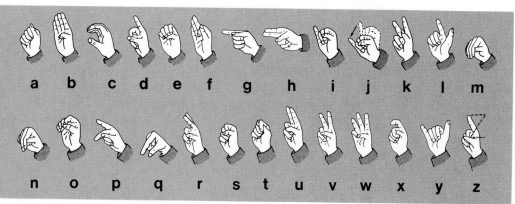

Sign language alphabet

Braille is a code of small, raised dots on paper that can be read by touch. People who are unable to see are able to read using Braille. And computers are made that have Braille dots on the keys and a machine that says the words as the blind person types them into the computer.

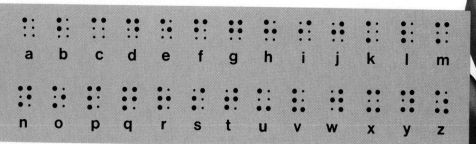

Braille alphabet

TRY THIS Activity!
What Would It Be Like?

What if you were blind or deaf? This activity will allow you to experience what it would be like if you couldn't see or hear.

What You Need
blindfold, *Activity Log* page 31

With a partner, blindfold yourself and have your partner lead you around the classroom. Remember to concentrate on where you're going. Have your partner lead you around so you don't run into anything. After a few minutes, blindfold your partner and lead him or her around.

What was it like not being able to see? Did you always know what was in front of you? In your *Activity Log* write a paragraph about the things you noticed when you were blindfolded.

Now, what's it like not being able to hear? Put your hands over your ears. Walk around the room and talk to your classmates. Try to communicate with them.

What did you experience about not being able to hear? How was it difficult to communicate? In your *Activity Log*, write down the ways you were able to communicate with your classmates.

Literature Link
Helen Keller—A Light for the Blind

Read the book by Kathleen V. Kudlinski called *Helen Keller—A Light for the Blind*, about a girl who was deaf as well as blind. Helen eventually learned how to communicate.

After reading this book, list all the things Helen wasn't able to do. Then list things she could do because of what she was taught. What methods did she use to communicate and be heard?

Helen Keller and her teacher

BY KATHLEEN V. KUDLINSKI
ILLUSTRATED BY DONNA DIAMOND

Help to Eyes and Ears

Danger to Your Eyes

Can light be hazardous to your eyes? Your eyes, like your skin, can be sunburned if you spend too much time in the bright sun or a tanning booth without protective eye-wear. Eyes are highly sensitive to ultraviolet (UV) rays. Injury can occur to the cornea, retina, and other parts of the eye within hours of being exposed. UV radiation is believed to cause cataracts (kat´ ə rakts), an eye problem that destroys eyesight.

Do you think people are aware of the damage UV rays might cause to their eyes? Investigate tanning booths and find out if customers are required to wear special glasses. In your *Activity Log* on page 32, write an article to the public, warning them about the damage UV light can cause and what they can do to prevent it.

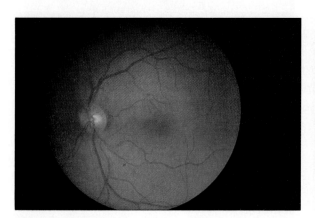

This is a picture of a healthy eye.

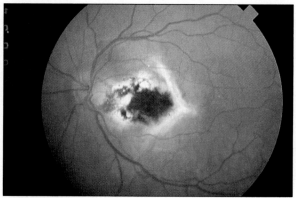

This is a picture of an eye damaged by UV radiation.

Health  Link

Assistance for the Deaf and the Blind

Sometimes illness or injury causes the body structures that interact with light or sound to break down. The result is blindness or deafness. New technology, combined with our knowledge of how we see and hear, has led to many devices that assist deaf and blind people in their daily lives.

Closed-captioned television enables deaf people to watch television and know what is being said. Flashing lights connected to alarm systems, and alarm clocks that vibrate the bed are other aids for the hearing impaired.

For the blind and visually challenged, there are devices that magnify television images, "talking" clocks and calculators, and robot guide dogs, like the one in the picture.

A waterproof hearing aid can help the hearing impaired communicate more easily.

Sum It Up

Your eyes and ears are complex structures that work every minute you're awake, catching light waves and sound vibrations from the air. Vibrations interacting with the small parts of your ear pass along information about what kind of sound you are hearing.

The amount of light that enters your eye is regulated by the pupil. You see color all around you because the lens in your eye interacts with the retina which refracts light.

Your brain plays an important part in seeing and hearing. It must decode the messages sent by the structures of your ears and eyes and tell you what you are hearing and seeing. Your sight and hearing are very valuable. In a normal day you hear and see without realizing that you are doing it.

Critical Thinking

1. What careers would be very difficult for someone with colorblindness?

2. What would happen if the small bones in your ear were damaged or missing?

3. What parts of the body's system work together to enable you to see and hear?

How Does Technology Extend Our Senses?

In the time of ancient civilizations, people could see only as far as the horizon and hear only those sounds close enough to reach their ears directly. Today, our eyes can see microscopic organisms and distant stars. And our ears can listen in on conversations halfway around the world.

Sound and light are energy. Together, they give you information about many things in your environment every day. The more we learn about sound and light, the more we can extend what we hear and see.

Our eyes can see into outer space through telescopes.

You know when you hear a stereo playing or see this book, it is because of the way sound and light waves interact with matter.

Through your activities, you have learned that because more energy goes into some vibrations than others, sounds are different. Because light waves bend when they hit obstacles, a dazzling array of color surrounds you. Although light travels faster than sound, it can't travel through a closed door like sound can.

As our understanding of sound and light increases, technology continually extends our ability to use the energy in sound and light to communicate, to make work easier, and to entertain.

Because of technology, we can see things we never would have guessed existed.

Video camera used to make home movies

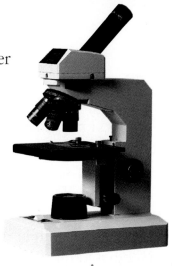

Microscopes using convex lenses help us examine tiny plants, animals, and body cells. Magnifying the images allows us to see what the human eye couldn't otherwise see.

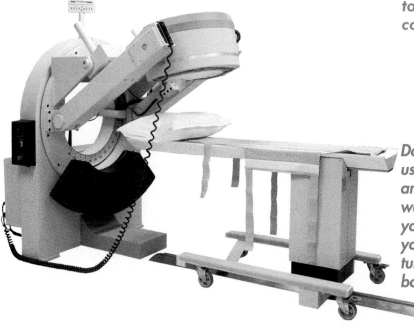

Doctors can examine your body using different frequencies of light and sound waves. X rays use waves that pass right through your body to make a picture of your bones. Ultrasound is used to turn echoes into pictures of the body.

Literature Link
Rainbows to Lasers and Sound Waves to Music

Look through the two books *Rainbows to Lasers* by Kathryn Whyman and *Sound Waves to Music* by Neil Ardley to find some information about sound and light that you haven't learned in this unit. In your **Activity Log** on page 33, write a short report about the information. Then present your report to the class.

Lasers

Science has found a way to use light to carry sound. Do you know that the music you hear on compact discs is reproduced by tiny lasers?

Lasers are light waves, perfectly in line like soldiers marching along, all in a straight line, and all at the same speed. Some lasers are powerful enough to cut steel. Others are delicate enough to clean teeth and fill cavities in "painless" dentistry. Lasers also read grocery prices at the check-out counter, and help detectives find hard-to-see fingerprints. Laser lights are used to make holograms, three-dimensional pictures that look like solid objects. A different kind of picture you may have seen is made by giant "dancing" laser lights at a music concert or fireworks display. These aren't holograms—they are very rapidly-moving laser beams that make outline pictures.

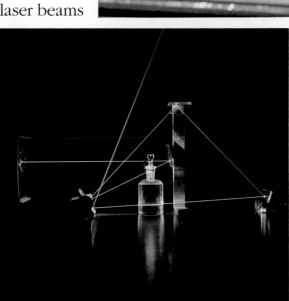

Lasers are very powerful, uniform beams of light that are changing the way many things are done in research, medicine, entertainment, communications, and industry.

Communication Around the World

Have you ever heard someone talk about the world "shrinking"? What they mean is that faster communication is turning Earth into one big neighborhood. Satellites above Earth send live television programs around the world as easily as across town, pass along telephone conversations from New York to Hong Kong, and monitor hurricanes in oceans, as part of worldwide weather forecasting. Even space is shrinking as space vehicles send pictures to Earth from such distant planets as Mars, Jupiter, Saturn, and Uranus.

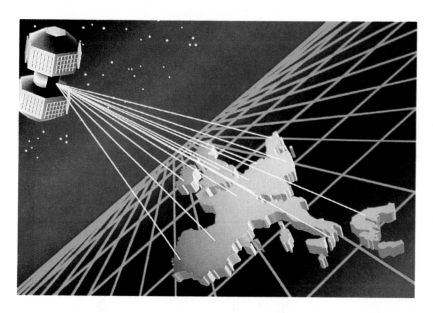

As scientists find new ways to use light and sound, communication between distant points on Earth gets faster and easier.

Music/Art  Link

A Room of Sound and Light

Using old magazines, newspapers, or catalogs, cut out furniture and household furnishings that you like. Pick flooring, lighting, wall-hangings, and entertainment equipment. On a large piece of paper, arrange and glue the pictures to make a model living room. How many colors of the spectrum does your room include? Would your room be dangerous to your eyes or ears? How would sound in this room affect your ability to sleep, read, or play? How would light affect these activities? Compare your model room to your classmates'. Which room would be the quietest? Which room would be the brightest? The noisiest? The darkest?

GLOSSARY

Use the pronunciation key below to help you decode, or read, the pronunciations.

Pronunciation Key

a	at, bad	d	dear, soda, bad
ā	ape, pain, day, break	f	five, defend, leaf, off, cough, elephant
ä	father, car, heart	g	game, ago, fog, egg
âr	care, pair, bear, their, where	h	hat, ahead
e	end, pet, said, heaven, friend	hw	white, whether, which
ē	equal, me, feet, team, piece, key	j	joke, enjoy, gem, page, edge
i	it, big, English, hymn	k	kite, bakery, seek, tack, cat
ī	ice, fine, lie, my	l	lid, sailor, feel, ball, allow
îr	ear, deer, here, pierce	m	man, family, dream
o	odd, hot, watch	n	not, final, pan, knife
ō	old, oat, toe, low	ng	long, singer, pink
ô	coffee, all, taught, law, fought	p	pail, repair, soap, happy
ôr	order, fork, horse, story, pour	r	ride, parent, wear, more, marry
oi	oil, toy	s	sit, aside, pets, cent, pass
ou	out, now	sh	shoe, washer, fish mission, nation
u	up, mud, love, double	t	tag, pretend, fat, button, dressed
ū	use, mule, cue, feud, few	th	thin, panther, both
ü	rule, true, food	th	this, mother, smooth
ù	put, wood, should	v	very, favor, wave
ûr	burn, hurry, term, bird, word, courage	w	wet, weather, reward
ə	about, taken, pencil, lemon, circus	y	yes, onion
b	bat, above, job	z	zoo, lazy, jazz, rose, dogs, houses
ch	chin, such, match	zh	vision, treasure, seizure

absorbed (ab sôrbd′) the act or process of being soaked up; to take in without reflection (light) or echo (sound).

auditory nerve (ô′di tôr′ ē nûrv) the nerve that carries impulses from the inner ear to the brain, so the brain receives the message that some sound reached the ears.

bioluminescence (bī ō′ lü mə nes′ əns) a giving off of light by a living organism, such as fireflies.

cochlea (kok′ lē ə) the tube of the inner ear, shaped somewhat like a snail shell, containing the sensory ends of the auditory nerve.

concave lens (kon′ kāv) a lens that is thinner in the middle than it is at the edges, and curves inward.

convex lens (kon veks′) a lens that is thicker in the middle than it is at the edges, and curves outward.

echo (ek′ ō) the repetition of a sound made when sound waves reflect from a distant surface.

frequency (frē′ kwən sē) the number of cycles per second of an alternating current, electromagnetic radiation, or sound.

image a picture of an object when light rays from the object are focused on a surface, as by a lens or mirror

laser a device that produces an extremely powerful beam of light consisting of light waves that are of the same wavelength, and whose waves vibrate in the same direction.

lens a piece of glass or other transparent material.

Mesopotamia (mes'ə pə ta' mē ə) a historic region in southwest Asia, between the Tigris and Euphrates rivers, a center of ancient civilization.

opaque (ō pāk') material that does not allow light to pass through.

pigments (pig' məntz) colored material that absorbs certain colors of light and reflects other colors.

pitch the way a person perceives the frequency of a sound, whether it's high or low.

pointillism (poin' til izm) the style of painting in which the artist applies the paint as numerous small dots.

prism (priz' əm) an object of transparent material with three straight faces at an angle to each other which reflects visible light into a spectrum.

pupil the opening in the center of the iris, through which light enters the eye.

reflect bouncing of a wave or ray off a surface, changing its direction.

refraction (ri frak'shən) the bending of waves when they change speed while passing from one material to another at an angle.

retina (ret' ə nə) the inner membrane at the back of the eyeball, made up of several layers of cells that are sensitive to light and transmit the images entering the eye to the optic nerve.

sonar (sō'när) a way of locating underwater objects by means of sound waves reflected from or produced by the objects.

spectrum (spek' trəm) a band of colors into which white light is separated according to wavelength, by being passed through a prism or other material.

translucent (trans lü' sənt) material that allows some light to pass through, but that doesn't allow objects on the other side to be clearly seen.

transparent (trans pâr' ənt) material that allows light to pass through, so that objects on the other side can be clearly seen.

ultrasonic (ul' trə son' ik) sound waves with frequencies higher than the range or limits of human hearing.

ultraviolet (ul' trə vī' ə lit) light waves that are part of the electromagnetic spectrum and can't be seen by the eyes; they have wavelengths shorter than those of visible light but longer than those of X rays.

vibration (vī brā' shən) a back and forth or side to side movement.

visible light the only part of the electromagnetic spectrum seen by the human eye.

volume (vol' ūm) perceived loudness of a sound, determined by the amplitude of a sound wave.

waves the regular disturbances caused by bumping molecules.

INDEX

CREDITS